3 Florida FAST Grade 4 Math Practice Tests

Full-Length Test Prep with Detailed Answer Explanations

Dr. A. Nazari

Published by View Math Education

ViewMath.com

Grade 4 Math: 3 Practice Tests

Hey there, quick checker! 

Three tests. That's all it takes to **see where you stand** and **build your confidence** fast.

- Short and focused — perfect for a quick review.
- Find your strengths in just three rounds.
- Fast results, real progress!

Ready for a quick check? Let's go! 

> **Three quick tests and you'll know exactly what to study next!**

📋 How to Use This Book 📋

Three tests, one clear plan — here's how to make them count.

📖 What's Inside

- **3 Practice Tests** — *quick, focused tests covering all Grade 4 topics*
- **Answer Key** — *detailed answers and explanations at the back*
- **Reference Pages** — *math symbols and multiplication table you can use during tests*
- **My Test Tracker** — *record your scores and see your growth at a glance*

📋 How to Take a Practice Test

1. **Set up:** *Find a quiet spot. Gather pencils, eraser, and scratch paper.*
2. **Time it:** *Set a timer (or ask a grown-up to). Try to finish without rushing.*
3. **Work through it:** *Answer every question. Skip hard ones and come back later.*
4. **Check your work:** *Use any extra time to double-check your answers.*
5. **Score it:** *Use the Answer Key. Write your score in the Test Tracker.*

☑ Multiple Choice

Pick the best answer from the choices given. Read ALL options before choosing — the first one that looks right isn't always the best!

✏ Short Answer

Solve the problem and write your answer. Show your work — even if you get the answer wrong, partial credit counts in many real tests!

> ℹ **Quick Tip:** *With only 3 tests, make each one count. Review every missed question before moving to the next test!*

Test-Taking Tips

Quick strategies to help you ace all 3 tests!

Before You Start

- Get a good night's sleep and eat a healthy snack.
- Have your supplies ready: pencils, eraser, scratch paper.
- Take three slow, deep breaths — you've got this!
- Remember: just 3 tests, so give each one your full focus.

During the Test

1. **Read each question twice.** Underline key words like "how many more," "product," or "estimate."
2. **Show your work.** Write out each step — it helps you catch mistakes.
3. **Use scratch paper.** Line up digits carefully for multi-digit problems.
4. **Skip and return.** Stuck on a question? Star it and move on. Come back with fresh eyes.
5. **Check with estimation.** Does your answer make sense? A quick estimate can catch big errors.
6. **Use your time wisely.** Don't spend too long on one problem.

Multiple Choice Tricks

- Read **all** choices first.
- Cross out answers you know are wrong.
- Plug your answer back into the problem.
- When in doubt, eliminate and guess — never leave it blank!

Watch Out For...

- Rushing through without reading carefully.
- Mixing up $\times$ and $+$ in word problems.
- Forgetting to regroup when adding or subtracting.
- Not simplifying fractions when asked.
- Skipping the "check your work" step.

Three tests is all you need for a quick check-up! Focus on each one, learn from your mistakes, and you'll be amazed how much you grow.

 What You'll Need 

Pencils

#2 pencils are perfect!

Eraser

Clean erasing = cleaner work

Scratch Paper

For all your work!

Quiet Space

Like a real test room

Timer (Optional)

Practice pacing yourself

Confidence!

Three tests, three wins!

✔ You CAN Use

- Pencils and eraser
- Scratch paper
- The reference pages in this book
- A ruler (for measurement questions)

✖ You CANNOT Use

- Calculator
- Phone or tablet
- Other books or notes
- Help from anyone else

👥 For Parents & Teachers

This compact 3-test set is ideal for a quick skills check. Simulate real test conditions when possible. After each test, review missed questions together and focus on **growth** ("You improved on fractions this time!") rather than the raw score.

X^1 Math Reference Sheet X^1

Symbol	Name	Meaning
$+$	Plus	Combine amounts
$-$	Minus	Find the difference
$\times$	Times	Multiply (repeated groups)
$\div$	Divide	Split into equal parts
$=$	Equals	Same value on both sides
$>$ $<$	Greater / Less Than	Compares two values
$\frac{a}{b}$	Fraction	a parts out of b equal parts
$.$	Decimal Point	Separates wholes from parts
$\angle$	Angle	Measured in degrees

Key Math Words

- **Factor** — a number you multiply
- **Product** — answer from multiplying
- **Quotient** — answer from dividing
- **Remainder** — left over after dividing
- **Numerator** — top of a fraction
- **Denominator** — bottom of a fraction
- **Equivalent** — equal in value
- **Perimeter** — distance around
- **Area** — space inside
- **Estimate** — a close, rounded guess

🔍 Word Problem Clue Words

Add (+)	*in all, altogether, total, combined, sum, increase*
Subtract (−)	*how many more, how many fewer, difference, left, remain*
Multiply (×)	*each, every, per, times as many, groups of, product*
Divide (÷)	*split equally, shared among, divided into, per group*

Find more at
ViewMath.com/FL-Grade4

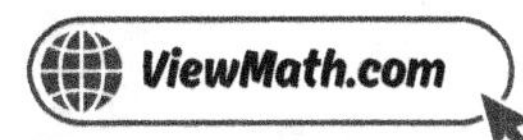

You may use this table during your practice tests!

×	1	2	3	4	5	6	7	8	9	10	11	12
1	1	2	3	4	5	6	7	8	9	10	11	12
2	2	4	6	8	10	12	14	16	18	20	22	24
3	3	6	9	12	15	18	21	24	27	30	33	36
4	4	8	12	16	20	24	28	32	36	40	44	48
5	5	10	15	20	25	30	35	40	45	50	55	60
6	6	12	18	24	30	36	42	48	54	60	66	72
7	7	14	21	28	35	42	49	56	63	70	77	84
8	8	16	24	32	40	48	56	64	72	80	88	96
9	9	18	27	36	45	54	63	72	81	90	99	108
10	10	20	30	40	50	60	70	80	90	100	110	120
11	11	22	33	44	55	66	77	88	99	110	121	132
12	12	24	36	48	60	72	84	96	108	120	132	144

💡 Quick Reminders

Multiply: Find one number on the left, the other on top. Where they meet = your answer.

Divide: For 96 ÷ 8, find 96 in the 8-row. The column header tells you the answer: 12!

Factor pairs: Any number in the table can be written as row header × column header.

📈 My Test Tracker 📈

Track your 3 tests and watch yourself improve!

Name: _______________________________________

Test #	Date	Score	How I Feel
1			
2			
3			

 *Three tests, three chances to shine! Compare each score to your last one — even one extra point means you learned something new!*

Quick progress is still progress! Three tests gave you a clear picture. Now you know exactly where to focus next. Be proud of every step forward!

 Table of Contents

Here's what we'll explore together!

 Let's learn and have fun!

Practice Test 1

30 Questions

✏️ Before You Start ✏️

- ✔ **Read each question carefully** before choosing your answer.
- ✔ **Show your work** on scratch paper when you need to.
- ✔ **Skip hard questions** and come back to them later.
- ✔ **Check your answers** when you're done.
- ✔ **Take your time** — there's no rush!

⭐ You've Got This! ⭐

Do your best and show what you know!

1. A box has 5 pencils. Another box has 7 times as many pencils. How many pencils are in the bigger box?

Your Answer:

2. There are 45 students split equally into 5 teams. Each team needs 3 jump ropes. How many jump ropes are needed in total?

Your Answer:

3. Is 100 a multiple of 4? Is 100 a multiple of 6? Answer both questions.

Your Answer:

4. Find the rule and fill in the missing value in the table:

In	Out
3	15
4	20
6	30
9	?

Your Answer:

5. In 307,416, which digit is in the hundred-thousands place?

(A) 7

(B) 4

(C) 3

(D) 0

Find more at
ViewMath.com/FL-Grade4

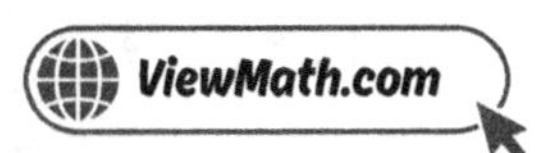

6. Which comparison is **INCORRECT?**

(A) $745,000 > 74,500$

(B) $600,001 > 599,999$

(C) $410,300 < 410,030$

(D) $325,400 = 325,400$

7. If $3 \times 10 = 30$, what is 3×100?

(A) 30

(B) 300

(C) 3,000

(D) 3,030

8. What is $1,000 + 999$?

(A) 1,899

(B) 1,909

(C) 1,999

(D) 2,099

9. List the partial products for 7×428 and find the total.

Your Answer:

10. A school collected 5,387 cans for a food drive. They want to pack them in bags of 7. How many full bags of 7 cans can they make?

(A) 769

(B) 769 R 4

(C) 770

(D) 768

11. Which fraction is the greatest?

(A) $\frac{1}{6}$

(B) $\frac{1}{3}$

(C) $\frac{1}{2}$

(D) $\frac{1}{4}$

Find more at
ViewMath.com/FL-Grade4

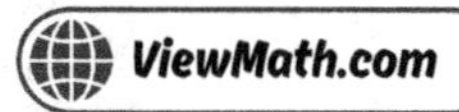

12. A pan of brownies is cut into 8 equal pieces. Leo ate $\frac{1}{8}$, Sam ate $\frac{2}{8}$, and Kai ate $\frac{1}{8}$. What fraction of the brownies did they eat altogether?

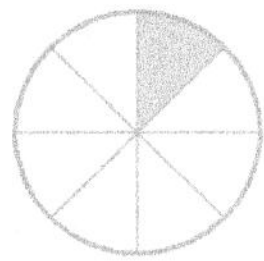

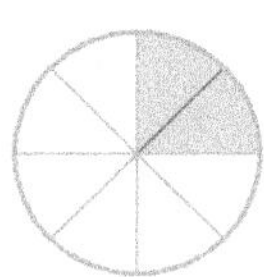

(A) $\frac{3}{8}$

(B) $\frac{4}{8}$

(C) $\frac{5}{8}$

(D) $\frac{6}{8}$

13. What is $\frac{5}{12} - \frac{2}{12}$?

(A) $\frac{3}{0}$

(B) $\frac{7}{12}$

(C) $\frac{3}{12}$

(D) $\frac{2}{12}$

14. What is $6\frac{3}{5} - 4\frac{4}{5}$? (Regroup as needed.)

(A) $2\frac{1}{5}$

(B) $1\frac{4}{5}$

(C) $2\frac{4}{5}$

(D) $1\frac{3}{5}$

15. What is $6 \times \frac{1}{4}$, written as a mixed number?

(A) $1\frac{1}{4}$

(B) $1\frac{3}{4}$

(C) $1\frac{2}{4}$

(D) $2\frac{1}{4}$

Find more at
ViewMath.com/FL-Grade4

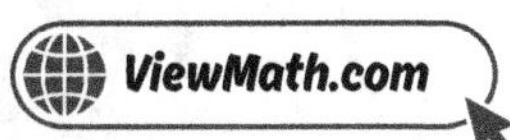

16. A dime is $\frac{1}{10}$ of a dollar, and a penny is $\frac{1}{100}$ of a dollar. How much of a dollar is 3 dimes and 8 pennies combined?

   (A) $\frac{30}{100}$

   (B) $\frac{38}{100}$

   (C) $\frac{11}{100}$

   (D) $\frac{308}{100}$

17. Write 0.25 as a fraction.

   (A) $\frac{25}{10}$

   (B) $\frac{25}{100}$

   (C) $\frac{2}{5}$

   (D) $\frac{2}{10}$

18. Which conversion is correct?

   (A) 1 cm = 10 mm

   (B) 1 m = 10 cm

   (C) 1 km = 100 m

   (D) 1 kg = 100 g

19. How many inches are in 1 foot?

   (A) 10

   (B) 12

   (C) 16

   (D) 3

Find more at
ViewMath.com/FL-Grade4

ViewMath.com

20. Look at the shopping receipt below. Find the total cost and the change from $20.00.

> **STORE RECEIPT**
>
> Notebook $3.50
> Pen pack $4.25
> Eraser $1.75
>
> **TOTAL: $?**
> **Change from $20: $?**

Your Answer:

21. A square room has sides of 10 feet. What is the area of the room?

(A) 40 sq ft

(B) 20 sq ft

(C) 100 sq ft

(D) 1,000 sq ft

22. What is the perimeter of a rectangle that is 9 cm long and 4 cm wide?

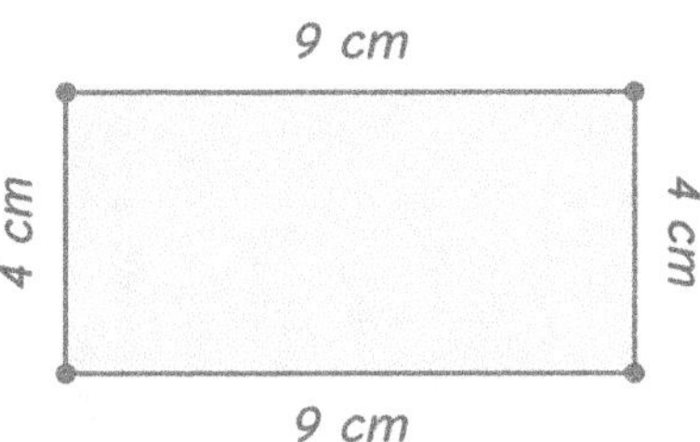

(A) 36 cm

(B) 26 cm

(C) 13 cm

(D) 52 cm

23. A number line for a line plot measures in $\frac{1}{2}$-cup intervals. Which values should appear on the number line between 0 and 1?

(A) $\frac{1}{4}$

(B) $\frac{3}{4}$

(C) $\frac{1}{2}$

(D) $\frac{2}{3}$

24. How many 90° angles make a full rotation of 360°?

(A) 2

(B) 3

(C) 4

(D) 6

25. The inner scale of a standard protractor starts at 0° on which side?

(A) Top

(B) Right

(C) Left

(D) Bottom

26. Four angles meet at a point and together make a full 360° turn. Three of the angles measure 90°, 90°, and 80°. What is the fourth angle?

(A) 80°

(B) 90°

(C) 100°

(D) 110°

27. How many points does it take to name a line?

(A) 1

(B) 2

(C) 3

(D) 4

Find more at
ViewMath.com/FL-Grade4

28. *True or false: Parallel lines can intersect if they are made long enough.*

Your Answer

29. *A square is BEST described as:*

(A) *A rectangle with 4 equal sides*

(B) *A rhombus with no right angles*

(C) *A trapezoid with 4 equal sides*

(D) *A triangle with 4 right angles*

30. *How many lines of symmetry does an equilateral triangle (all sides equal) have?*

(A) 1

(B) 2

(C) 3

(D) 0

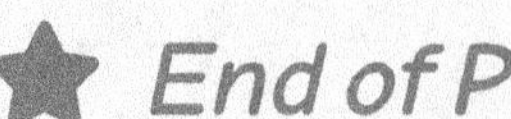 

⭐ *End of Practice Test 1* ⭐

Great job finishing the test!

 **My Score**

I got ____________ out of 30 questions right.

*Check your answers in the **Answer Key** at the back of the book.*

💡 Review any questions you missed. That's how we learn!

📊 Check Your Score Online!

*Visit **ViewMath Academy** to enter your answers and see which topics you need to review. You can also explore lessons, take quizzes, track your scores, and save your progress!*

viewmath.com/score/4.1.FL.01

Or go to viewmath.com/score and enter code: 4.1.FL.01

2

Practice Test 2

📋 *30 Questions*

✏️ Before You Start ✏️

- ✔ **Read each question carefully** before choosing your answer.
- ✔ **Show your work** on scratch paper when you need to.
- ✔ **Skip hard questions** and come back to them later.
- ✔ **Check your answers** when you're done.
- ✔ **Take your time** — there's no rush!

⭐ You've Got This! ⭐

Do your best and show what you know!

1. How many total dots are in the array below? Write a multiplication equation for the array.

9 columns

3 rows

Your Answer:

2. A school ordered 7 boxes of crayons. Each box has 24 crayons. The teacher gave away 15 crayons. How many crayons does the school have now?

(A) 168

(B) 153

(C) 163

(D) 152

3. I am a prime number between 45 and 55. What am I? (There may be more than one answer.)

Your Answer:

4. A pattern starts at 3 and the rule is "multiply by 3." What is the fourth term?

(A) 12

(B) 27

(C) 81

(D) 9

5. What is the value of the digit 5 in 850,293?

(A) 5

(B) 500

(C) 5,000

(D) 50,000

  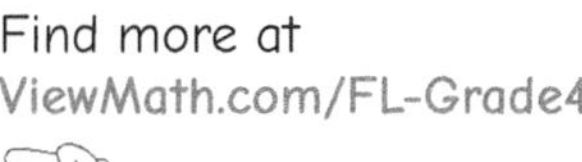

Find more at
ViewMath.com/FL-Grade4

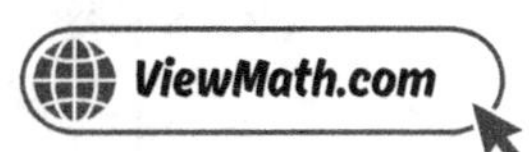

6. The population of Town A is 183,400 and Town B is 183,040. Which town has more people?

(A) Town B

(B) They are equal

(C) Town A

(D) Cannot tell

7. Look at the diagram below. The same digit 9 appears in two positions. How many times bigger is the value at Position A than at Position B?

TTh	Th	H	T	O
9	2	**9**	4	1

↓ A   ↓ B

Your Answer:

8. What is $432 + 251$?

(A) 673

(B) 683

(C) 693

(D) 783

9. Each page of a photo album holds 7 photos. The album has 215 pages. How many photos can the album hold?

Your Answer:

10. $3,200 \div 8 =$ __________

(A) 40

(B) 400

(C) 4,000

(D) 40,000

Find more at
ViewMath.com/FL-Grade4

11. *True or false:* $\frac{2}{3} > \frac{5}{6}$.

Your Answer

12. *Which is a correct way to decompose $\frac{7}{8}$?*

(A) $\frac{4}{8} + \frac{4}{8}$

(B) $\frac{3}{8} + \frac{4}{8}$

(C) $\frac{5}{8} + \frac{3}{8}$

(D) $\frac{2}{8} + \frac{6}{8}$

13. *Which addition has the same result as $\frac{5}{8} - \frac{2}{8}$?*

(A) $\frac{1}{8} + \frac{1}{8}$

(B) $\frac{1}{8} + \frac{2}{8}$

(C) $\frac{2}{8} + \frac{1}{8} + \frac{1}{8}$

(D) $\frac{2}{8} + \frac{2}{8}$

14. *What is $3\frac{7}{8} - 1\frac{5}{8}$?*

(A) $2\frac{3}{8}$

(B) $1\frac{7}{8}$

(C) $2\frac{2}{8}$

(D) $1\frac{2}{8}$

Find more at
ViewMath.com/FL-Grade4

15. What is $5 \times \frac{1}{3}$, written as a mixed number?

(A) $\frac{5}{3}$

(B) $1\frac{2}{3}$

(C) $2\frac{1}{3}$

(D) $1\frac{1}{3}$

16. What fraction must be added to $\frac{3}{10}$ to get $\frac{45}{100}$?

(A) $\frac{15}{100}$

(B) $\frac{42}{100}$

(C) $\frac{5}{100}$

(D) $\frac{75}{100}$

17. Look at the 10×10 grid below. What decimal represents the shaded part?

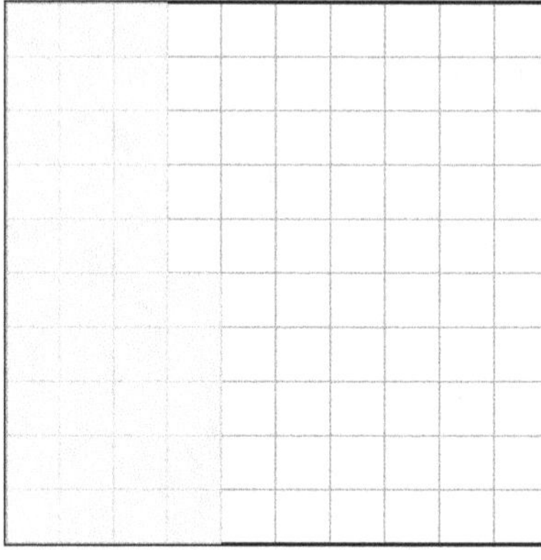

(A) 0.035

(B) 0.35

(C) 3.5

(D) 35.0

18. How many centimeters are in 2 kilometers? (two-step: km → m → cm)

(A) 2,000 cm

(B) 20,000 cm

(C) 200,000 cm

(D) 2,000,000 cm

19. Look at the balance scale. The left side has 5 pounds. How many ounces are needed on the right to balance it?

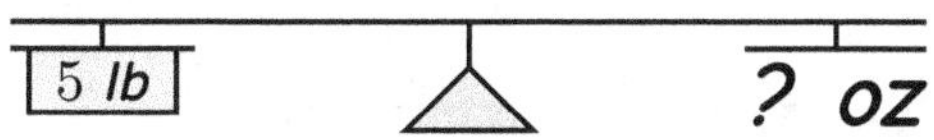

(A) 60 oz

(B) 80 oz

(C) 50 oz

(D) 90 oz

20. Look at the timeline below. A movie starts at 1:50 PM and is 2 hours 25 minutes long. What time does it end? Show your work on the timeline.

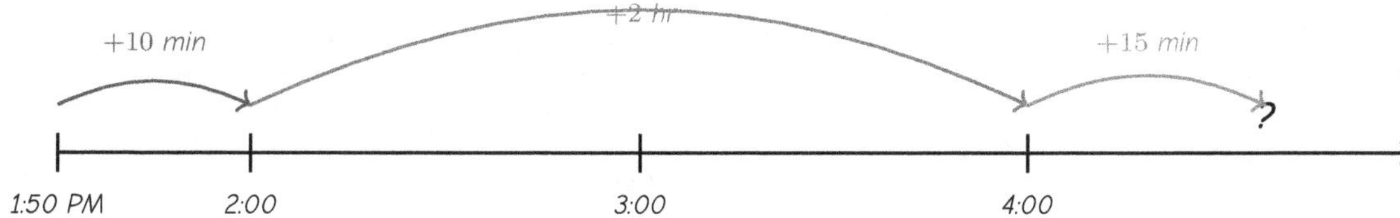

Your Answer:

21. A classroom floor is 9 m × 8 m. A bookshelf covers a 2 m × 3 m area. What area of floor is NOT covered by the bookshelf?

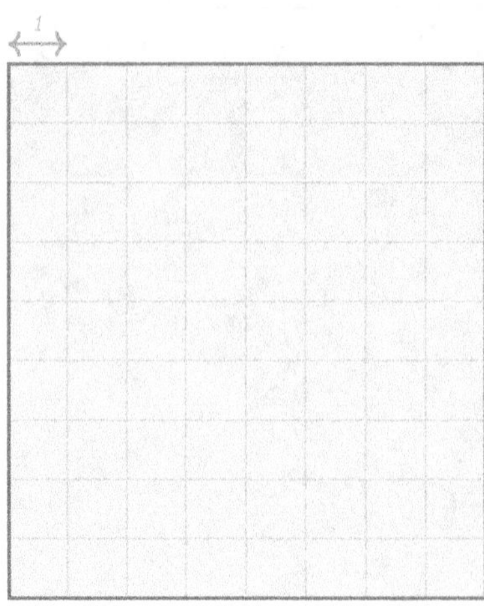

(A) 66 sq m

(B) 54 sq m

(C) 48 sq m

(D) 60 sq m

22. Look at the shape below made from two connected rectangles. What is the total perimeter?

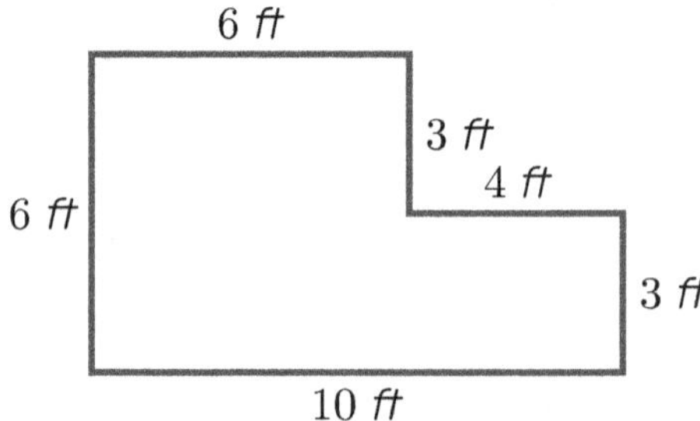

(A) 28 ft

(B) 30 ft

(C) 32 ft

(D) 36 ft

23. Lisa measures pencil lengths: 2 pencils at $\frac{5}{8}$ in, 3 pencils at $\frac{6}{8}$ in, 1 pencil at $\frac{7}{8}$ in. What is the longest pencil length measured?

(A) $\frac{5}{8}$ in

(B) $\frac{6}{8}$ in

(C) $\frac{7}{8}$ in

(D) $\frac{8}{8}$ in

24. An angle that measures 135° is best described as:

(A) Acute

(B) Right

(C) Obtuse

(D) Straight

25. When drawing a 75° angle using a protractor, after placing one ray along the baseline, you make a dot at which reading on the scale?

(A) 75°

(B) 105°

(C) 15°

(D) 90°

26. Three angles sit on a straight line (180°). Two of them measure 60° and 70°. What is the third angle?

(A) 130°

(B) 50°

(C) 60°

(D) 40°

27. Look at the figure. Name the line segment shown using its endpoints.

P   Q

Your Answer:

28. Which of these real-world examples best shows perpendicular lines?

    (A) Two lanes of a highway going straight ahead   (B) The top and side edges of a door

    (C) Two train tracks running side by side   (D) The two sides of a ladder

29. How many sides does a quadrilateral have?

    (A) 3   (B) 4

    (C) 5   (D) 6

30. How many lines of symmetry does an equilateral triangle have?

    Your Answer:

 # End of Practice Test 2 

Great job finishing the test!

☑ My Score

I got __________ out of 30 questions right.

*Check your answers in the **Answer Key** at the back of the book.*

💡 Review any questions you missed. That's how we learn!

📊 Check Your Score Online!

Visit **ViewMath Academy** to enter your answers and see which topics you need to review. You can also explore lessons, take quizzes, track your scores, and save your progress!

viewmath.com/score/4.1.FL.02

Or go to viewmath.com/score and enter code: 4.1.FL.02

Practice Test 3

 30 Questions

 Before You Start

- ✓ **Read each question carefully** before choosing your answer.
- ✓ **Show your work** on scratch paper when you need to.
- ✓ **Skip hard questions** and come back to them later.
- ✓ **Check your answers** when you're done.
- ✓ **Take your time** — there's no rush!

⭐ You've Got This! ⭐

Do your best and show what you know!

1. The number bond below shows a multiplication comparison. What number belongs in the top circle?

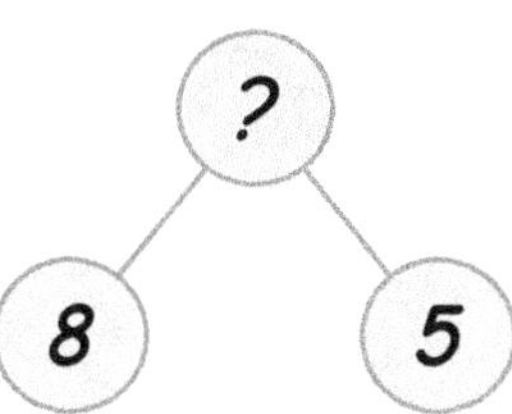

Hint: The top number is 8 times as many as 5.

Your Answer

2. Sofia baked 48 cookies. She put them equally into 6 bags. She then bought 5 bags from a friend. How many bags does she have altogether?

(A) 11

(B) 8

(C) 13

(D) 43

3. Which statement is TRUE about all even numbers greater than 2?

(A) They are all prime.

(B) They are all composite.

(C) They are neither prime nor composite.

(D) Some are prime and some are composite.

4. Create your own number pattern using the rule "Add 8, starting at 4." Write the first five terms.

Your Answer

5. Look at the base-ten blocks below. What number do they show?

(A) 374

(B) 347

(C) 743

(D) 437

6. A stadium holds 72,500 fans. A second stadium holds 72,050 fans. Which stadium holds fewer fans?

(A) First stadium

(B) Second stadium

(C) They hold equal fans

(D) Cannot determine

7. Look at the place value chart below. The digit 7 moves from one position to another.

	TTh	Th	H	T	O
Before:	0	0	7	0	0
After:	7	0	0	0	0

How did the value of the digit 7 change?

(A) It became 10 times bigger

(B) It became 100 times bigger

(C) It became 10 times smaller

(D) It stayed the same

8. What is $5,478 + 3,865$?

Your Answer:

9. *What is 7×905?*

(A) 6,235

(B) 6,325

(C) 6,335

(D) 6,435

10. *A factory makes 9,764 widgets. They pack them in boxes of 6. How many full boxes are made?*

(A) 1,627

(B) 1,628

(C) 1,626

(D) 1,629

11. *A student says $\frac{5}{6} > \frac{7}{8}$ because $5 < 7$. Is the student correct?*

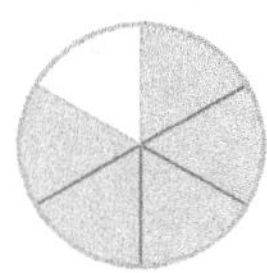

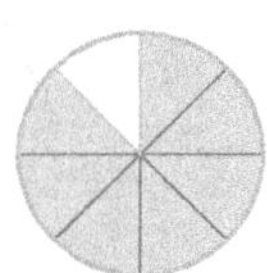

(A) *Yes, because 5 is less than 7.*

(B) *No. $\frac{5}{6} < \frac{7}{8}$ because $\frac{5}{6} \approx 0.83$ and $\frac{7}{8} \approx 0.875$.*

(C) *Yes, because $6 < 8$.*

(D) *No. $\frac{5}{6} = \frac{7}{8}$.*

12. $\frac{6}{10}$ can be broken into $\frac{4}{10} +$ _____.

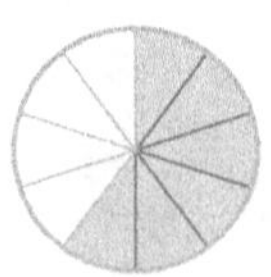

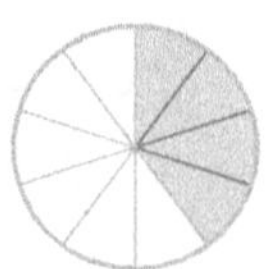

   (A) $\frac{3}{10}$ (B) $\frac{2}{10}$

   (C) $\frac{4}{10}$ (D) $\frac{6}{10}$

13. What is $\frac{4}{6} - \frac{1}{6}$?

   (A) $\frac{3}{12}$ (B) $\frac{4}{6}$

   (C) $\frac{3}{0}$ (D) $\frac{3}{6}$

14. What is $1\frac{2}{6} + 2\frac{3}{6}$?

   (A) $3\frac{1}{6}$ (B) $3\frac{5}{12}$

   (C) $4\frac{2}{6}$ (D) $3\frac{5}{6}$

15. Look at the number line below. It shows 4 equal jumps of $\frac{3}{6}$ starting from 0. Where does the last arrow land?

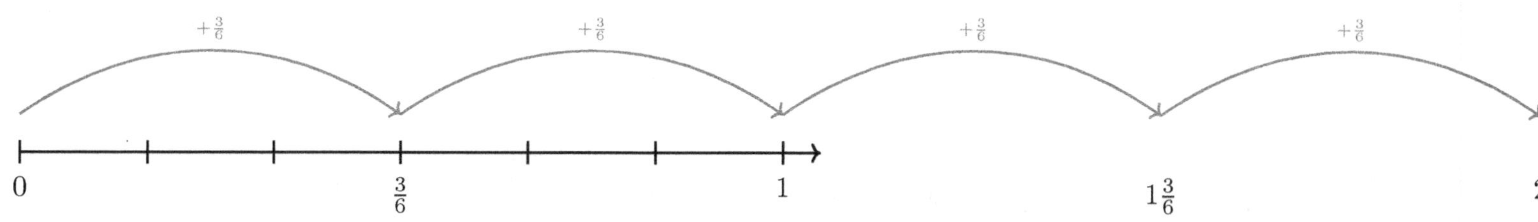

(A) $1\frac{2}{6}$ (B) $1\frac{3}{6}$

(C) 2   (D) $2\frac{3}{6}$

16. What is $\frac{2}{10} + \frac{35}{100}$?

Your Answer:

17. A snack costs $0.48. Write this price as a fraction of a dollar in two ways: once with a denominator of 100, and once simplified.

Your Answer:

18. How many centimeters are in 1 meter?

(A) 10   (B) 100

(C) 1,000   (D) 10,000

19. Convert 36 inches to feet.

(A) 2 ft   (B) 3 ft

(C) 4 ft   (D) 12 ft

Find more at
ViewMath.com/FL-Grade4

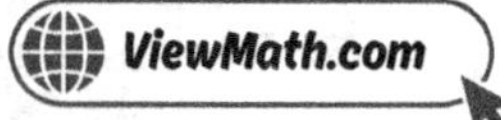

20. Rosa buys a book for $6.75 and a pencil for $0.49. How much does she spend in total?

(A) $6.24

(B) $7.24

(C) $7.14

(D) $7.34

21. Look at the two rectangles below. What is their combined area?

Your Answer:

22. A square has a perimeter of 48 inches. What is the length of one side?

(A) 4 in

(B) 8 in

(C) 12 in

(D) 24 in

23. A line plot shows: 2 X's at $\frac{1}{4}$, 3 X's at $\frac{1}{2}$, and 1 X at $\frac{3}{4}$. How many total data points are there?

(A) 3

(B) 6

(C) 8

(D) 12

24. Estimate: An angle opens slightly more than a right angle but much less than a straight angle. What benchmark angles is it between? Give an estimated measure.

Your Answer:

25. An acute angle reads 72° on the inner scale and 108° on the outer scale. What is the correct measurement?

Your Answer:

26. When an angle is split into two smaller parts, the whole angle equals the ___________ of its parts.

(A) difference

(B) product

(C) sum

(D) quotient

27. Which of the following is NOT true about a point?

(A) It marks an exact location

(B) It has no size

(C) It is drawn as a small dot

(D) It can be measured in inches

28. Which pair of lines below is **perpendicular**?

A.

B.

C.

(A) Pair A

(B) Pair B

(C) Pair C

(D) Both Pair A and Pair C

29. A quadrilateral with **exactly one pair** of parallel sides is called a:

(A) Rectangle

(B) Parallelogram

(C) Trapezoid

(D) Rhombus

30. How many lines of symmetry does a square have?

Your Answer:

Find more at
ViewMath.com/FL-Grade4

 End of Practice Test 3 

Great job finishing the test!

 My Score

I got _____________ out of 30 questions right.

Check your answers in the **Answer Key** at the back of the book.

💡 Review any questions you missed. That's how we learn!

📊 Check Your Score Online!

Visit **ViewMath Academy** to enter your answers and see which topics you need to review. You can also explore lessons, take quizzes, track your scores, and save your progress!

viewmath.com/score/4.1.FL.03

Or go to *viewmath.com/score* and enter code: 4.1.FL.03

Answer Key & Explanations

⭐ Check Your Answers! ⭐

First try each test on your own, then look here to check.
Read the explanations to learn from any mistakes ⭐

✅ Practice Test 1 — Answer Key

 **1** 35

 **2** 15

3 *Yes, 100 is a multiple of 4* $(4 \times 25 = 100)$. *No, 100 is not a multiple of 6* $(100 \div 6 = 16\ R4)$.

4 *Rule: Multiply by 5; missing value = 45*

 **5** *C*    **6** *C*    **7** *B*    **8** *C*    **9** 2,996

 **10** *A*    **11** *C*    **12** *B*    **13** *C*    **14** *B*    **15** *C*    **16** *B*    **17** *B*    **18** *A*   **19** *B*

20 *Total = \$9.50; change = \$10.50*    **21** *C*    **22** *B*    **23** *C*    **24** *C*    **25** *C*    **26** *C*

 **27** *B*    **28** False    **29** *A*   **30** *C*

💡 Time to Learn! 💡

*Go through the explanations below, **especially for the questions you missed**.*

Understanding why each answer is correct makes you a stronger math thinker!

👍 **Tip:** *Circle any questions you got wrong, then read their explanation carefully.*

📖 Practice Test 1 — Detailed Explanations

 Find more at
ViewMath.com/FL-Grade4
☞

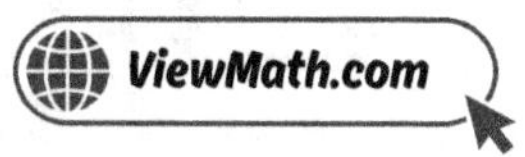

1. $7 \times 5 = 35$ pencils.

2. Step 1: $45 \div 5 = 9$ students per team (not needed). There are 5 teams each needing 3 jump ropes: $5 \times 3 = 15$.

3. $100 \div 4 = 25$ exactly; $100 \div 6$ leaves a remainder.

4. $3 \times 5 = 15$, $4 \times 5 = 20$, $6 \times 5 = 30$, so $9 \times 5 = 45$.

5. The hundred-thousands place is the leftmost digit of a six-digit number. In 307,416, the digit 3 is in the hundred-thousands place.

6. In 410,300 vs 410,030: compare hundreds: $3 > 0$. So $410,300 > 410,030$. The $<$ sign is wrong—this comparison is INCORRECT.

7. $3 \times 100 = 300$. Multiplying by 100 shifts all digits two places to the left (same as multiplying by 10 twice).

8. $1,000 + 999 = 1,999$. You can see this because 999 is 1 less than 1,000, so the total is 1 less than 2,000.

9. $7 \times 400 = 2,800$; $7 \times 20 = 140$; $7 \times 8 = 56$. Total: $2,800 + 140 + 56 = 2,996$.

10. $7 \times 769 = 5,383$. $5,387 - 5,383 = 4$. So $5,387 \div 7 = 769$ R 4. There are 769 full bags.

11. All have the same numerator (1). The smaller the denominator, the larger the fraction. $\frac{1}{2}$ has the smallest denominator, so it is greatest.

12. $\frac{1}{8} + \frac{2}{8} + \frac{1}{8} = \frac{1+2+1}{8} = \frac{4}{8}$.

13. $5 - 2 = 3$; keep denominator 12: $\frac{3}{12}$.

14 Regroup: $6\frac{3}{5} = 5\frac{8}{5}$. Then $5 - 4 = 1$ and $\frac{8}{5} - \frac{4}{5} = \frac{4}{5}$. Answer: $1\frac{4}{5}$.

15 $6 \times 1 = 6$; $\frac{6}{4} = 1\frac{2}{4}$ because $6 \div 4 = 1 \text{ R } 2$.

16 3 dimes $= \frac{3}{10} = \frac{30}{100}$. Add 8 pennies $= \frac{8}{100}$. Total: $\frac{30}{100} + \frac{8}{100} = \frac{38}{100}$.

17 0.25 is 25 hundredths $= \frac{25}{100}$.

18 $1 \text{ cm} = 10 \text{ mm}$. The other conversions are wrong: $1 \text{ m} = 100 \text{ cm}$, $1 \text{ km} = 1{,}000 \text{ m}$, $1 \text{ kg} = 1{,}000 \text{ g}$.

19 There are 12 inches in 1 foot.

20 $\$3.50 + \$4.25 + \$1.75 = \$9.50$. Change: $\$20.00 - \$9.50 = \$10.50$.

21 $A = 10 \times 10 = 100$ sq ft.

22 $P = 2(9) + 2(4) = 18 + 8 = 26$ cm.

23 When measuring in halves, the only value between 0 and 1 is $\frac{1}{2}$.

24 $360° \div 90° = 4$.

25 The inner scale begins at $0°$ on the **left** side of the protractor and increases toward the right.

26 $90 + 90 + 80 = 260$. Fourth angle $= 360 - 260 = 100°$.

27 Any two points on a line can be used to name it, such as $\overleftrightarrow{AB}$.

Find more at
ViewMath.com/FL-Grade4

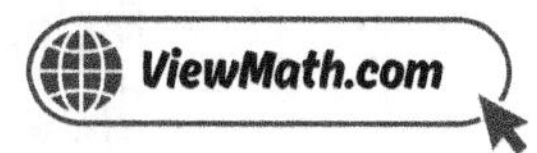

28 Parallel lines never intersect, no matter how far they are extended.

29 A square is a special rectangle (4 right angles) that also has all 4 sides equal. It is also a special rhombus.

30 An equilateral triangle has 3 lines of symmetry — one from each vertex to the midpoint of the opposite side.

✅ Practice Test 2 — Answer Key

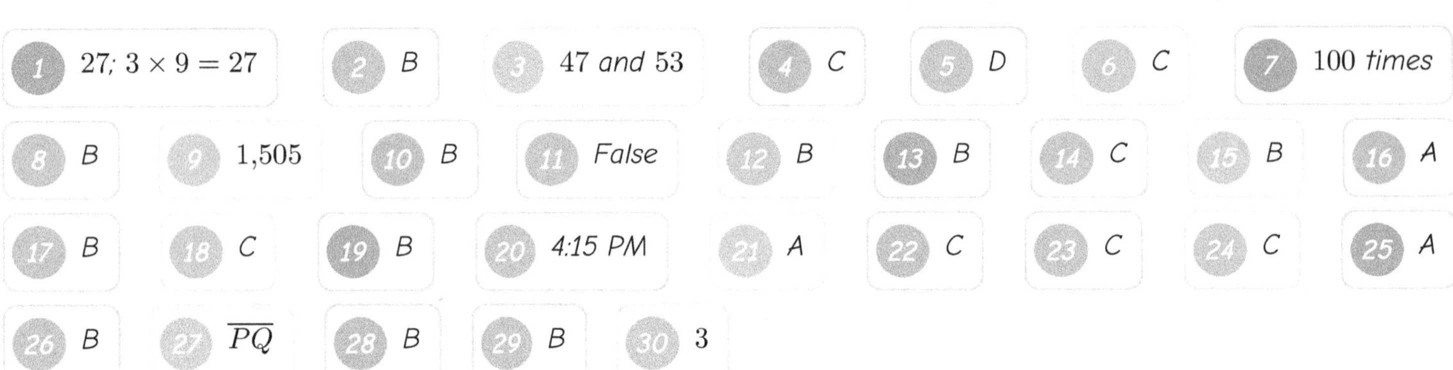

1	$27; 3 \times 9 = 27$	2	B	3	47 and 53	4	C	5	D	6	C	7	100 times
8	B	9	1,505	10	B	11	False	12	B	13	B	14	C	15	B	16	A
17	B	18	C	19	B	20	4:15 PM	21	A	22	C	23	C	24	C	25	A
26	B	27	$\overline{PQ}$	28	B	29	B	30	3

💡 Time to Learn! 💡

Go through the explanations below, **especially for the questions you missed**.

Understanding why each answer is correct makes you a stronger math thinker!

👍 **Tip:** Circle any questions you got wrong, then read their explanation carefully.

📖 Practice Test 2 — Detailed Explanations

1 The array has 3 rows and 9 columns. $3 \times 9 = 27$ dots total.

Find more at
ViewMath.com/FL-Grade4

2 Step 1: $7 \times 24 = 168$ crayons. Step 2: $168 - 15 = 153$ crayons.

3 47 has only factors 1 and 47. 53 has only factors 1 and 53. All other numbers between 45 and 55 are composite.

4 $3 \times 3 = 9$ (2nd), $9 \times 3 = 27$ (3rd), $27 \times 3 = 81$ (4th).

5 In 850,293: 8 = hundred-thousands, 5 = ten-thousands. The value is $5 \times 10,000 = 50,000$.

6 Both towns have 183 in the thousands period. Compare hundreds: $4 > 0$. So $183,400 > 183,040$. Town A has more people.

7 Position A is the ten-thousands place ($9 \times 10,000 = 90,000$). Position B is the hundreds place ($9 \times 100 = 900$). $90,000 \div 900 = 100$. Position A is 100 times bigger.

8 Adding column by column: ones $2 + 1 = 3$, tens $3 + 5 = 8$, hundreds $4 + 2 = 6$. The sum is 683.

9 7×215: ones $7 \times 5 = 35$, write 5, carry 3; tens $7 \times 1 + 3 = 10$, write 0, carry 1; hundreds $7 \times 2 + 1 = 15$. Product is 1,505.

10 $32 \div 8 = 4$, and $3,200 = 32 \times 100$, so $3,200 \div 8 = 4 \times 100 = 400$.

11 Convert $\frac{2}{3} = \frac{4}{6}$. Compare $\frac{4}{6}$ and $\frac{5}{6}$: $4 < 5$, so $\frac{2}{3} < \frac{5}{6}$.

12 The numerators must add to 7. Only $3 + 4 = 7$ works. ($4 + 4 = 8$, $5 + 3 = 8$, $2 + 6 = 8$ — all wrong.)

13 $\frac{5}{8} - \frac{2}{8} = \frac{3}{8}$. Check: $\frac{1}{8} + \frac{2}{8} = \frac{3}{8}$ ✓.

14 Whole numbers: $3 - 1 = 2$. Fractions: $\frac{7}{8} - \frac{5}{8} = \frac{2}{8}$. Answer: $2\frac{2}{8}$.

Find more at
ViewMath.com/FL-Grade4

15 $5 \times 1 = 5$; $\frac{5}{3} = 1\frac{2}{3}$ because $5 \div 3 = 1\ R\ 2$.

16 $\frac{3}{10} = \frac{30}{100}$. The difference: $\frac{45}{100} - \frac{30}{100} = \frac{15}{100}$.

17 35 squares out of 100 are shaded. $\frac{35}{100} = 0.35$.

18 Step 1: 2 km $= 2 \times 1,000 = 2,000$ m. Step 2: 2,000 m $= 2,000 \times 100 = 200,000$ cm.

19 1 lb $= 16$ oz. So 5 lb $= 5 \times 16 = 80$ oz.

20 From 1:50 PM $+10$ min $= 2:00$ PM. Then $+2$ hours $= 4:00$ PM. Then $+15$ min $= 4:15$ PM. ($10 + 15 = 25$ minutes total, plus 2 hours.)

21 Floor area $= 9 \times 8 = 72$ sq m. Bookshelf area $= 2 \times 3 = 6$ sq m. Remaining: $72 - 6 = 66$ sq m.

22 Add all outer sides: $10 + 3 + 4 + 3 + 6 + 6 = 32$ ft.

23 $\frac{7}{8}$ inch is the largest value on the line plot.

24 $135°$ is between $90°$ and $180°$, so it is an obtuse angle.

25 You mark a dot at exactly $75°$ on the correct scale, then draw a ray from the vertex to that dot.

26 $60 + 70 = 130$, so the third angle $= 180 - 130 = 50°$.

27 The figure shows a segment with endpoints P and Q, so it is line segment $\overline{PQ}$.

28 The top edge and side edge of a door meet at a $90°$ angle, making them perpendicular.

Find more at
ViewMath.com/FL-Grade4

29   *"Quad" means four. A quadrilateral always has exactly 4 sides and 4 angles.*

30   *An equilateral triangle has 3 lines of symmetry, one from each vertex to the midpoint of the opposite side.*

✅ Practice Test 3 — Answer Key

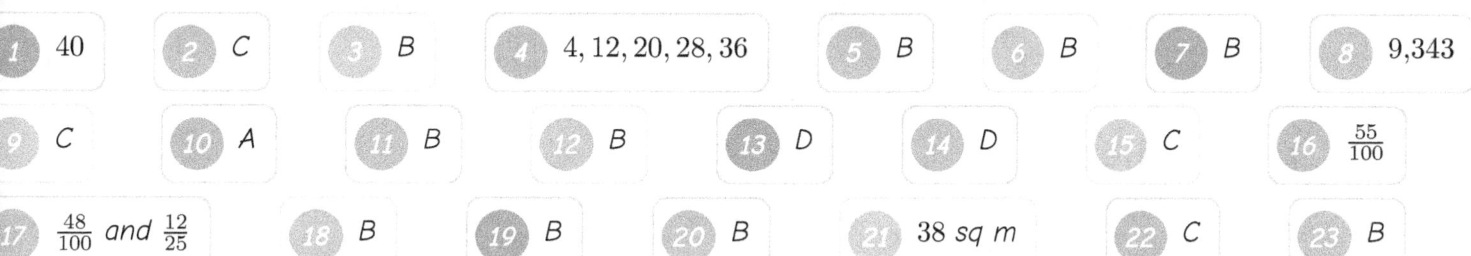

1 40	**2** C	**3** B	**4** 4, 12, 20, 28, 36	**5** B	**6** B	**7** B	**8** 9,343
9 C	**10** A	**11** B	**12** B	**13** D	**14** D	**15** C	**16** $\frac{55}{100}$
17 $\frac{48}{100}$ and $\frac{12}{25}$	**18** B	**19** B	**20** B	**21** 38 sq m	**22** C	**23** B	

24 Between 90° and 180°; obtuse; approximately 90°–180° (e.g., 120°)   **25** 72°   **26** C   **27** D

28 B   **29** C   **30** 4

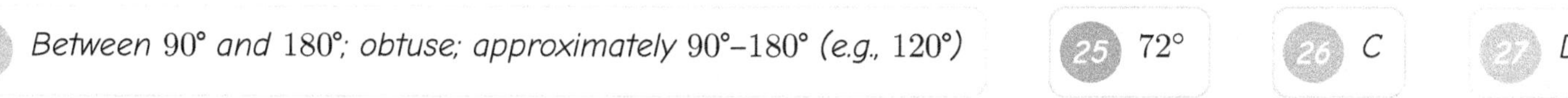

💡 Time to Learn! 💡

Go through the explanations below, **especially for the questions you missed**.

Understanding why each answer is correct makes you a stronger math thinker!

👍 **Tip:** Circle any questions you got wrong, then read their explanation carefully.

📖 Practice Test 3 — Detailed Explanations

1   *8 times as many as 5 is $8 \times 5 = 40$.*

2   *Step 1: $48 \div 6 = 8$ bags baked. Step 2: $8 + 5 = 13$ bags total.*

Find more at
ViewMath.com/FL-Grade4

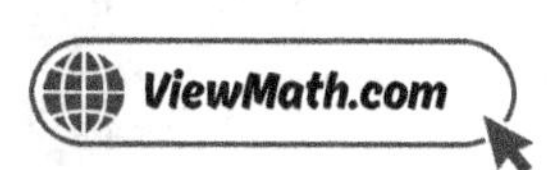

3   Every even number greater than 2 has 2 as a factor (in addition to 1 and itself), giving more than 2 factors, so all are composite.

4   $4 + 8 = 12$, $12 + 8 = 20$, $20 + 8 = 28$, $28 + 8 = 36$.

5   There are 3 hundreds (300), 4 tens (40), and 7 ones (7). That makes $300 + 40 + 7 = 347$.

6   Both start with 72. Compare hundreds: $5 > 0$. So $72{,}500 > 72{,}050$. The second stadium holds fewer fans.

7   The 7 moved from the hundreds place (700) to the ten-thousands place (70,000). That's two places to the left: $700 \times 10 \times 10 = 70{,}000$, which is 100 times bigger.

8   Ones: $8 + 5 = 13$, carry 1. Tens: $7 + 6 + 1 = 14$, carry 1. Hundreds: $4 + 8 + 1 = 13$, carry 1. Thousands: $5 + 3 + 1 = 9$. Sum is 9,343.

9   Ones: $7 \times 5 = 35$, write 5, carry 3. Tens: $7 \times 0 + 3 = 3$. Hundreds: $7 \times 9 = 63$. Product is 6,335.

10   $6 \times 1{,}627 = 9{,}762$. Remainder: $9{,}764 - 9{,}762 = 2$. So there are 1,627 full boxes with 2 extra widgets.

11   You cannot compare numerators when denominators differ. Using a common denominator of 24: $\frac{5}{6} = \frac{20}{24}$ and $\frac{7}{8} = \frac{21}{24}$. So $\frac{5}{6} < \frac{7}{8}$.

12   $4 + ? = 6$, so $? = 2$. The missing piece is $\frac{2}{10}$.

13   Subtract the numerators, keep the denominator. $4 - 1 = 3$, so $\frac{4}{6} - \frac{1}{6} = \frac{3}{6}$.

14   Whole numbers: $1 + 2 = 3$. Fractions: $\frac{2}{6} + \frac{3}{6} = \frac{5}{6}$. Answer: $3\frac{5}{6}$.

15   $4 \times \frac{3}{6} = \frac{12}{6} = 2$. Each jump is $\frac{3}{6}$ $(= \frac{1}{2})$, and 4 jumps of $\frac{1}{2}$ lands on 2.

Find more at
ViewMath.com/FL-Grade4

16 $\frac{2}{10} = \frac{20}{100}$. Then $\frac{20}{100} + \frac{35}{100} = \frac{55}{100}$.

17 $0.48 = \frac{48}{100}$. Simplified by dividing by 4: $\frac{48 \div 4}{100 \div 4} = \frac{12}{25}$.

18 There are 100 centimeters in 1 meter.

19 $36 \div 12 = 3$ ft.

20 $\$6.75 + \$0.49 = \$7.24$.

21 Rectangle A: $5 \times 4 = 20$ sq m. Rectangle B: $6 \times 3 = 18$ sq m. Total: $20 + 18 = 38$ sq m.

22 A square has 4 equal sides: $48 \div 4 = 12$ in per side.

23 $2 + 3 + 1 = 6$ total data points.

24 Since the angle is bigger than 90° but much less than 180°, it is obtuse. A reasonable estimate is around 120° or so.

25 Because the angle is acute (less than 90°), read the scale giving a value below 90°: 72°.

26 Angle measurement is additive: the whole angle equals the **sum** of its parts.

27 A point is simply a location in space. It has no size, so it cannot be measured in inches.

28 Perpendicular lines meet at a 90° angle (shown by the small square). Only Pair B has a right angle mark.

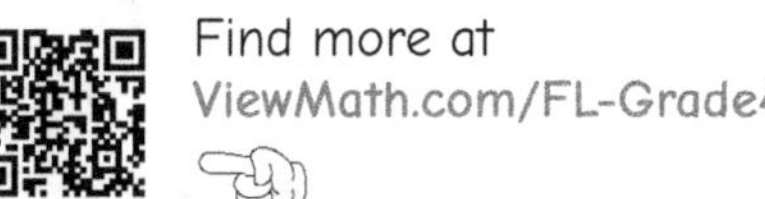

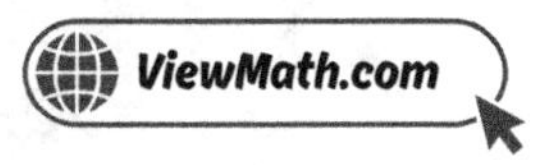

29 A trapezoid has exactly 1 pair of parallel sides. A parallelogram, rectangle, rhombus, and square all have 2 pairs.

30 A square has 4 lines of symmetry: one vertical, one horizontal, and two diagonal.

Great job checking your work!

Keep practicing and you'll be a math star!

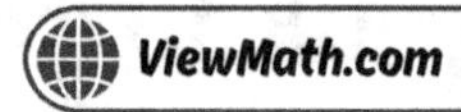

www.ingramcontent.com/pod-product-compliance
Lightning Source LLC
Chambersburg PA
CBHW081238130726
47997CB00009B/2910